Key Stage 2

Decimals and Percentages

Steve Mills and Hilary Koll

Name _______________________________

Schofield & Sims

Introduction

You might ask why you need to know about decimals and percentages. Well, they are both just other ways to write fractions, which are parts of things. A half is the same as 0·5 and also 50 per cent. When working with units of measurement we usually use decimals, and when looking at problems using large numbers we may find percentages to be more useful. In this learning workbook you will find information about how to write and use decimals and percentages. You will practise working with them, and learn when and how to use them.

How to use this book

Before you start using this book, write your name in the name box on the first page.

Then decide how to begin. If you want a complete course on decimals and percentages, you should work right through the book from beginning to end. Another way to use the book is to dip into it when you want to find out about a particular topic. The contents page will help you to find the pages you need.

Whichever way you choose, don't try to do too much at once – it's better to work through the book in short bursts.

When you have found the topic you want to study, look out for these icons, which mark different parts of the text:

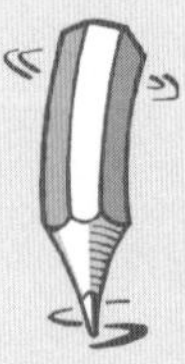

Activities

This icon shows you the activities that you should complete. You write your answers in the spaces provided. You might find it useful to have some scrap paper to work on for some of the activities. After you have worked through all the activities on the page, turn to pages A1 to A3 at the centre of the book to check your answers. When you are sure that you understand the topic, put a tick in the box beside it on the Contents page.

On pages 10 and 16, you will find **Progress Tests**. These contain questions that will check your understanding of the topics that you have worked through so far. Check your answers on page A4. It is important that you correct any mistakes before moving on to the next section.

At the back of the book you will find a **Final Test**. This will check your understanding of all the topics (page 26).

Explanation

This text explains the topic and gives examples. Make sure you read it before you start the activities.

Scrap Paper

This icon tells you when you may need to use scrap paper to work out your answers.

Fascinating Facts

This text gives you useful background information about the subject.

Decimals and Percentages

Schofield & Sims | Understanding Maths

Contents

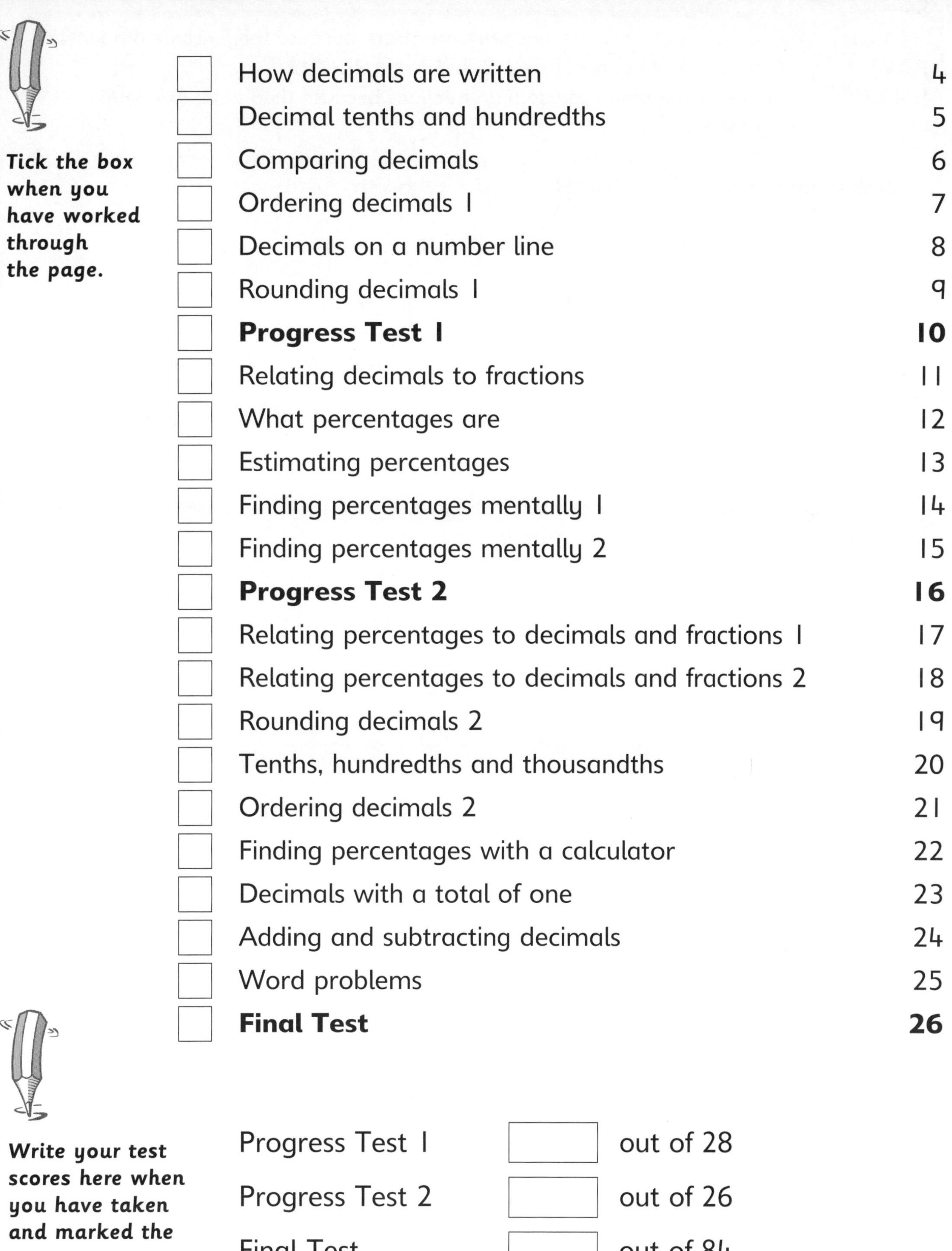

Tick the box when you have worked through the page.

Write your test scores here when you have taken and marked the tests.

Progress Test 1 ☐ out of 28

Progress Test 2 ☐ out of 26

Final Test ☐ out of 84

How decimals are written

Decimals, like fractions, are **part numbers** because they include amounts that are less than one, like **0·3**, **4·68** and **36·782**

Decimals are easier to use than fractions because they work like whole numbers.

With **whole numbers**...   Th H T U · t h th

10 units make **1** ten, **10** tens make **1** hundred, **10** hundreds make **1** thousand etc.

With **decimals**...   Th H T U · t h th

10 thousandths make **1** hundredth, **10** hundredths make **1** tenth, **10** tenths make **1** unit and so on.

When you are working with decimals you need to know what each digit in the number stands for...

tenths   hundredths   thousandths

T U · t h th

4 6 · 5 8 2

whole numbers   part numbers

The numbers to the left of the decimal point show how many whole numbers we have.	The numbers to the right show us how many tenths, hundredths, thousandths etc. we have.

One tenth means one out of ten 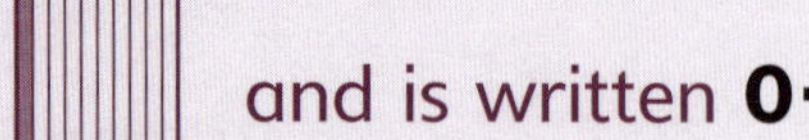 and is written **0·1**

One hundredth means one out of a hundred 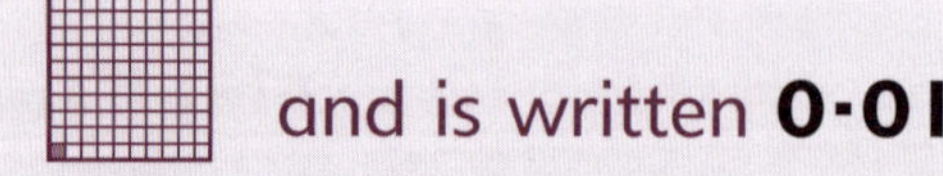 and is written **0·01**

So **0·34** means **3** tenths and **4** hundredths or **34** hundredths 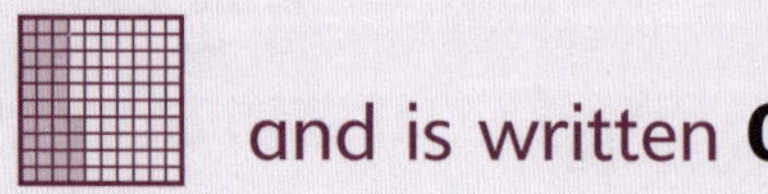 and is written **0·34**

Decimal tenths and hundredths

Did you know...? The **decimal point** separates whole numbers from part numbers.

$$3\,9\cdot4\,2$$

whole numbers | part numbers

decimal point

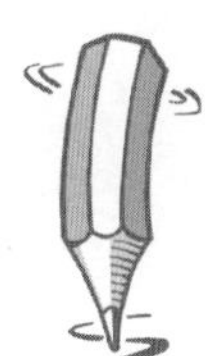

Tenths

The first digit to the right of the decimal point is the tenths digit. This is because **10** tenths make a whole one.

U	·	t	h
6	·	8	5

1. Draw a ring around the **tenths** digit in these numbers.

a) 2·6 b) 6·3 c) 7·9

d) 1 3·6 e) 2 8·1 f) 5 6·4

g) 8·5 4 h) 9·5 6 i) 2 3·6 3

Hundredths

The second digit to the right of the decimal point is the hundredths digit. **10** hundredths make **1** tenth and **100** hundredths make a whole one.

U	·	t	h
8	·	2	7

2. Draw a ring around the **hundredths** digit in these numbers.

a) 2·7 2 b) 5·4 3 c) 9·0 2

d) 2 7·5 1 e) 6 5·9 7 f) 1 2 8·4 6

g) 5·4 9 2 h) 8·0 7 2 i) 1 9·6 0 8

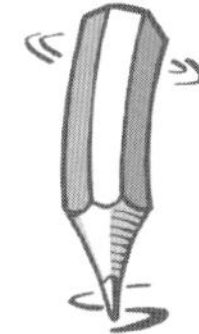

3. What is the pink digit worth in each of these numbers?

a) 2·56 _6 hundredths_ b) 18·42 __________ c) 38·97 __________

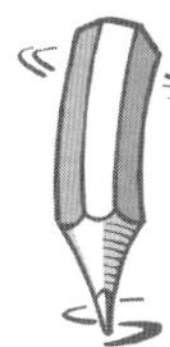

d) 156·64 __________ e) 280·85 __________ f) 56·841 __________

Comparing decimals

It's important to realise that tenths are **larger** than hundredths, and hundredths are **larger** than thousandths, and so on.

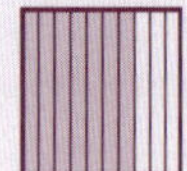 **0·7** is larger than **0·07** 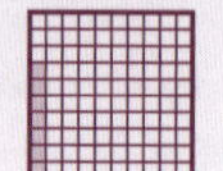

It's also important to know that decimal numbers with more digits after the decimal point are not always larger.

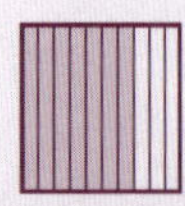 **0·7** is larger than **0·68** 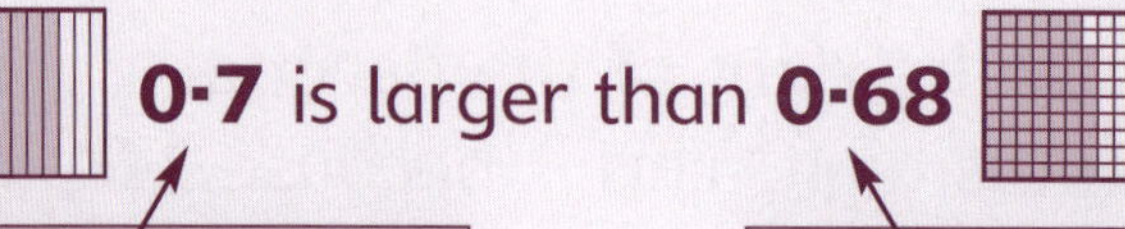

Think of this as **0·70** where the zero shows it has no hundredths. It is easier to compare **0·70** with **0·68** as they have the same number of digits after the decimal point.

This number has **2** digits after the decimal point but it does not mean it has to be larger. Always see how many tenths it has.

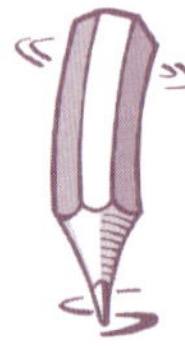

1. Draw a ring around the larger number in each pair.

a) 
0·6   0·67

b) 
0·5   0·48

c) 
0·77   0·8

d) 
0·6   0·61

e) 
0·72   0·7

f) 
0·9   0·89

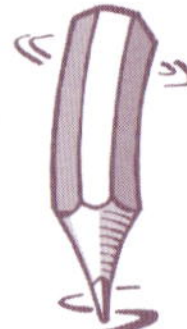

2. Write the numbers in the bags to make the scales tilt correctly.

a) 2·6kg, 2·8kg

b) 3·75kg, 3·57kg

c) 5·12kg, 5·21kg

d) 4·8kg, 4·79kg

e) 6·05kg, 6·1kg

f) 7·2kg, 7·18kg

Ordering decimals

Ordering decimals is just like ordering whole numbers. Start by comparing the digit on the left and, if they are the same, move to the right to compare the next digit.

Put these decimals in order of size, largest first

$$0·61 \quad 0·53 \quad 0·7 \quad 0·54$$

Which number has most **tenths**? ⟶ **0·7**

Which has the next greatest number of **tenths**? ⟶ **0·61**

Which numbers have the next greatest number of **tenths**? ⟶ **0·53** or **0·54**

Which of these has most **hundredths**? ⟶ **0·54**

Which is the smallest number? ⟶ **0·53**

From largest to smallest the order is: **0·7, 0·61, 0·54, 0·53**

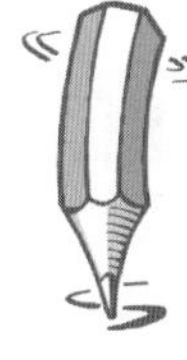

1. Put these decimals in order of size, **smallest** first.

a) 0·9, 0·8, 0·85, 0·56 _______________________________

b) 0·7, 0·75, 0·83, 0·8 _______________________________

c) 0·38, 0·41, 0·4, 0·39 _______________________________

d) 1·86, 1·9, 1·8, 1·87 _______________________________

e) 2·63, 1·98, 2·0, 2·07 _______________________________

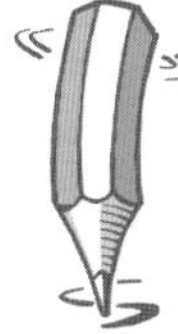

2. Write these in order of size, **largest** first.

a) 0·24kg, 0·42kg, 0·30kg, 0·4kg

b) 4·5m, 5·6m, 5·9m, 6·0m

c) £0.59, £0.50, £5.90, £9.50

d) 1·65kg, 6·15kg, 5·61kg, 6·2kg

Decimals on a number line

We can mark whole numbers on a number line like this

We can do the same with decimals.

This packet of sweets holds **10** sweets.

This picture shows **1·8** packets of sweets.
This means **1** whole packet and **8** tenths of a packet.

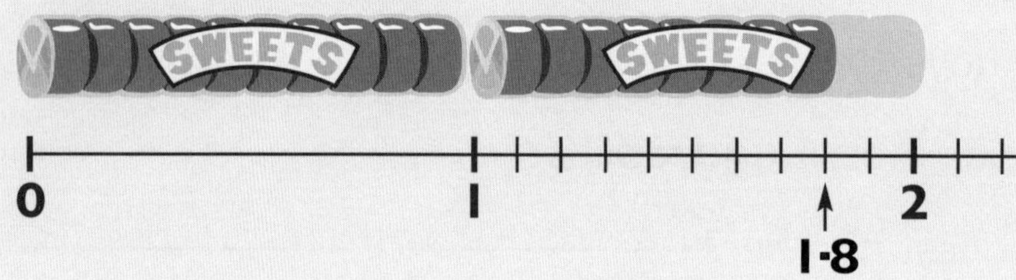

1·8

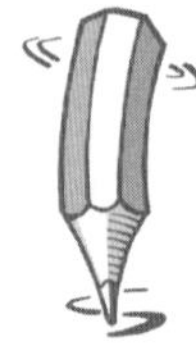

1. Mark how many packets of sweets are shown below.

a)
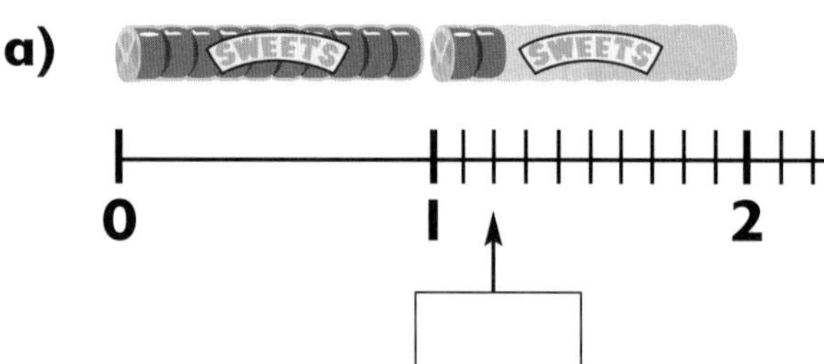

b)
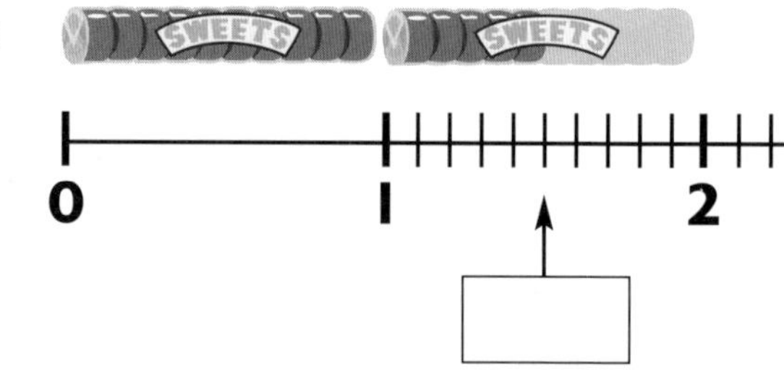

c)
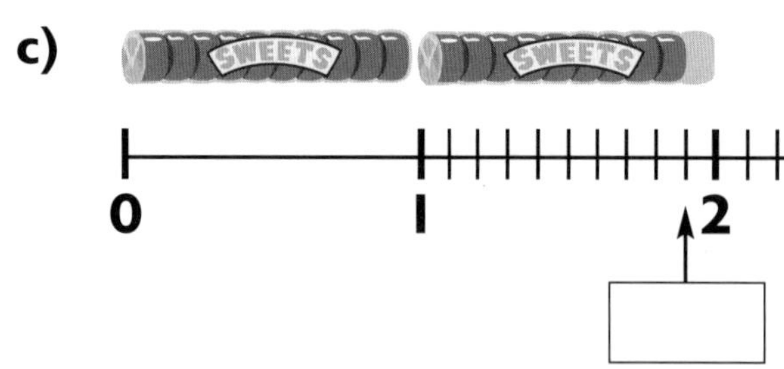

d)
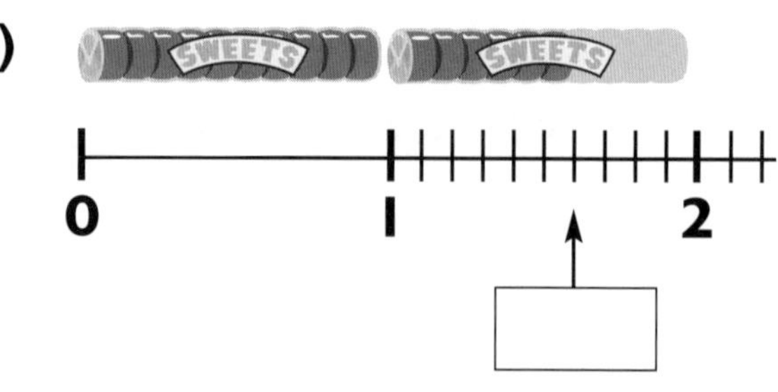

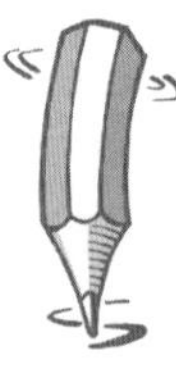

2. Mark these numbers on the number lines.

a) 1·6, 1·1, 0·7, 0·2

b) 1·4, 2·4, 0·9, 1·8

c) 3·5, 4·3, 4·8, 5·6

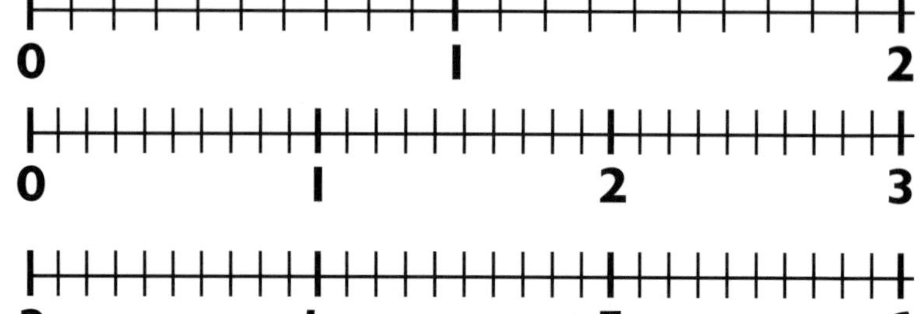

Rounding decimals 1

We round numbers to:
- give a good idea of a number – there were about **50 000** people at the United match
- give us a rough answer before we calculate: **23 × 18** is about **20 × 20 = 400**

Rounding decimals to the nearest whole number

We can round decimals in the same way we round whole numbers.

Look at this question: ***Round 7·3 to the nearest whole number.***

One way of rounding is to look at the number on a number line and see which whole number it is nearest to.

7   7·3   8

7·3 is nearest to **7**

Here is another way...

- Firstly see what you are rounding to.
  Round **7·3** to the nearest **whole number**

- Then look at this digit in the number...
  U · t h
  7 · 3

- Now look at the digit to the right of it....
  7 · 3

- If it's smaller than **5** the units digit stays the same.

- If it's **5** or larger the units digit must go up one.

  So **7·3** rounded to the nearest whole number is **7**

> **5 is in the middle. What do we do?**
>
> We round numbers like **3·5**, **4·5**, **9·5** up to the next whole number!

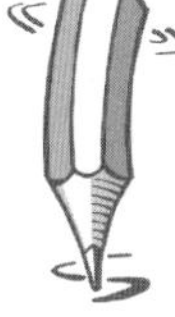

1. Round these numbers to the nearest **whole number**.

a) 3·7 →

b) 4·3 →

c) 5·8 →

d) 7·5 →

e) 10·5 →

f) 8·74 →

g) 12·06 →

h) 13·49 →

i) 20·09 →

j) 26·51 →

k) 31·199 →

l) 36·501 →

Progress Test 1

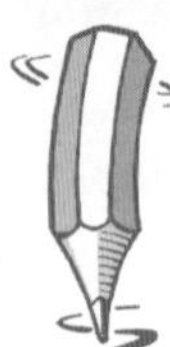

1. Draw a ring round the **tenths** digit in these numbers.

a) 7·6   b) 3·3 8   c) 1 7·9 5   d) 2 4·0 2 5

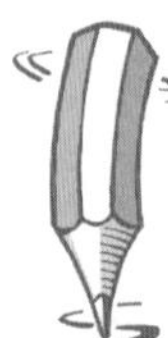

2. Draw a ring round the **hundredths** digit in these numbers.

a) 4·7 6   b) 1 5 4·7 3   c) 9 4·7 1 2   d) 2 0·0 2 4

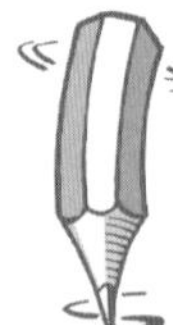

3. What is the **pink** digit worth in each of these numbers?

a) 4·59 __________   b) 24·43 __________   c) 78·216 __________

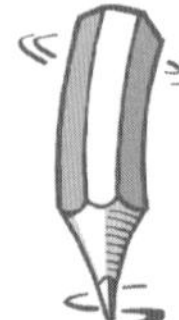

4. Draw a ring around the **larger** number in each pair.

a) 0·3   0·29   b) 0·7   0·79   c) 0·9   0·10

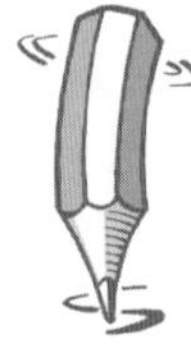

5. Put these decimals in order of size, **smallest** first.

a) 0·7, 0·63, 0·59, 0·66 _______________________________

b) 0·1, 0·2, 0·18, 0·81 _______________________________

c) 0·58m, 0·61m, 0·5m, 0·49m _______________________________

6. Mark these numbers on the number lines.

a) 1·8, 1·3, 0·9, 0·4

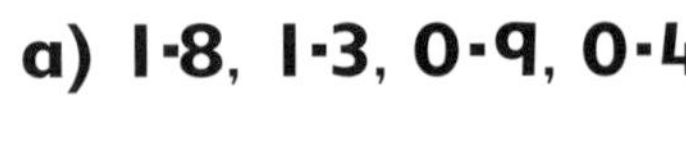

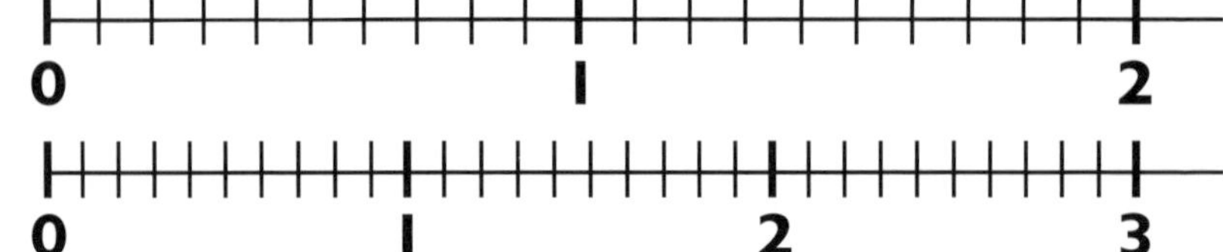

b) 0·7, 2·1, 1·9, 1·3

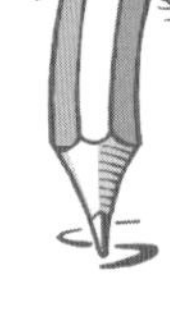

7. Round these numbers to the nearest **whole number.**

a) (3·6) → ( )   b) (8·39) → ( )   c) (25·492) → ( )

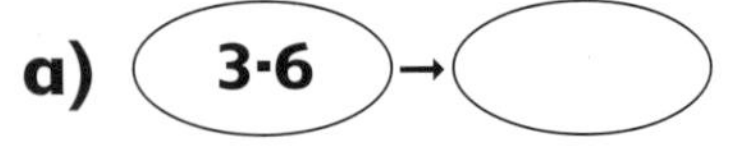
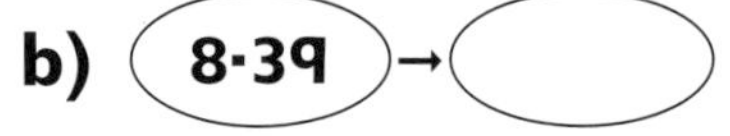
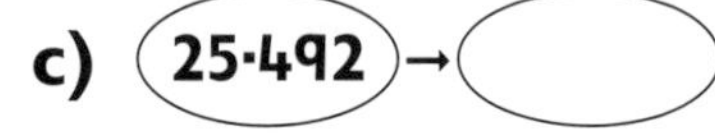

Decimals and fractions are like two different languages.

Different languages mean we can say the same thing in different ways.

French people and Greek people say 'Good morning' in different ways. But they mean the same thing.

Equivalence of decimals and fractions

Decimals and fractions are just different ways of saying the same thing.

A half, or **0·5**, is the same as the fraction $\frac{1}{2}$

A quarter, or **0·25**, is the same as the fraction $\frac{1}{4}$

Three quarters, or **0·75**, is the same as the fraction $\frac{3}{4}$

A tenth, or **0·1**, is the same as the fraction $\frac{1}{10}$

A hundredth, or **0·01**, is the same as the fraction $\frac{1}{100}$

So **0·43** means **4** tenths and **3** hundredths or **43** hundredths $\frac{43}{100}$

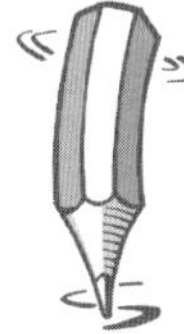

1. Can you move across the grid colouring in **pairs** of touching squares with the same value as each other? Squares can touch diagonally.

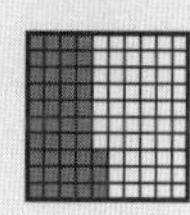

Start → ... → Finish

$\frac{1}{6}$	0.8	$\frac{3}{5}$	0.5	$\frac{6}{10}$	$\frac{1}{10}$	0.1	0.34	$\frac{8}{9}$	0.1
$\frac{1}{2}$	$\frac{1}{5}$	$\frac{3}{8}$	$\frac{4}{5}$	0.4	$\frac{2}{5}$	0.73	$\frac{15}{100}$	0.76	0.63
$\frac{2}{7}$	0.5	0.1	0.3	0.2	$\frac{2}{3}$	$\frac{3}{7}$	0.15	$\frac{63}{100}$	$\frac{8}{9}$
0.12	$\frac{1}{4}$	0.25	$\frac{3}{10}$	$\frac{1}{7}$	$\frac{87}{100}$	$\frac{3}{5}$	0.75	0.11	0.2

What percentages are

Percent means out of a hundred. **Per** means 'each' or 'every' and **cent** is a Roman word that means 'a hundred'.

Think about **cent**imetre – there's 100cm in a metre,

cent – there are 100 cents in an American dollar,

centipede – they're supposed to have 100 legs

and **cent**ury – 100 years.

The symbol that means 'per cent' is %. We can write **100**%, **35**% and **2**%.

We read these as 'one hundred per cent', 'thirty-five per cent' and 'two per cent'.

A **percentage** is a fraction with a **denominator**, or bottom number, of **100**, but written in a different way. **36**% means $\frac{36}{100}$.

We can write **100**%, which means the whole thing. Look on a shirt or T-shirt. The label might say '**100**% cotton' which means it is all cotton.

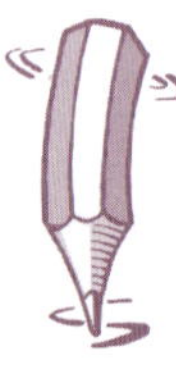

1. Complete these clothing labels, making sure they total **100**%.

a)

50%	cotton
_____	nylon

b)

20%	cotton
_____	silk

c)

_____	cotton
60%	wool

d)

40%	wool
_____	nylon
30%	silk

e)

45%	polyester
15%	cotton
_____	nylon

f)

75%	wool
_____	nylon
10%	cotton

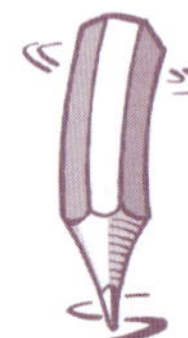

2. Explain what each poster means.

a)

"We will be trying **100**%" says player.

b)

'This yoghurt is **90**% fat free'.

c)

'**50**% off all cds'.

_________________________   _________________________   _________________________

_________________________   _________________________   _________________________

_________________________   _________________________   _________________________

Estimating percentages

Making an estimate is like having a good guess at something. Imagine a weather forecaster. She makes estimates all the time about how likely it is to rain or be sunny. She might say 'There's a **60%** chance of rain today'.

Don't worry that your estimates will be wrong. It doesn't matter if the weather forecaster wasn't exactly right – it was just an estimate to give people some idea of what to expect.

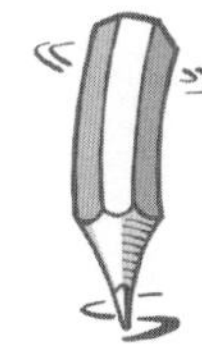

1. Match each container with an appropriate estimate.

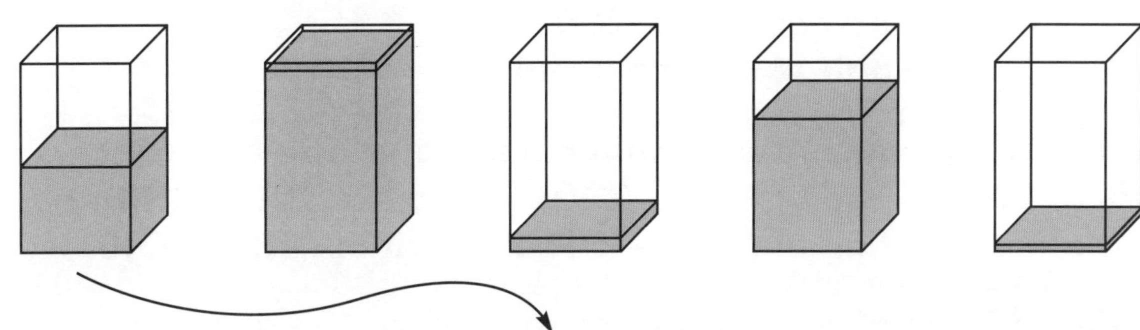

About 0% full	About 5% full	About 10% full	About 20% full	About 50% full	About 65% full	About 75% full	About 95% full	About 100% full

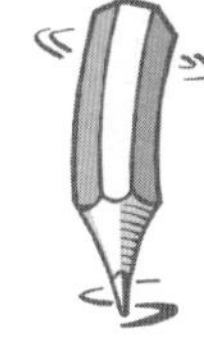

2. Match each shape with an appropriate estimate.

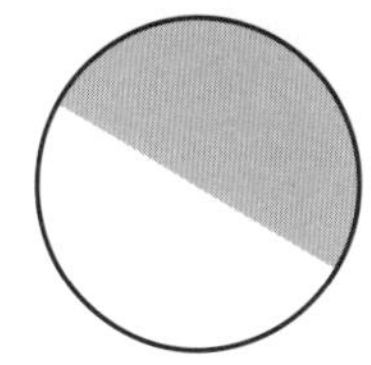 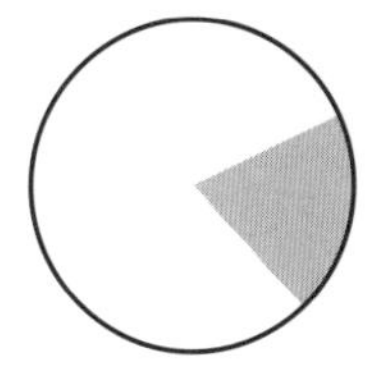 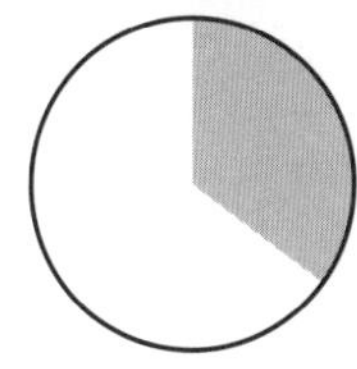 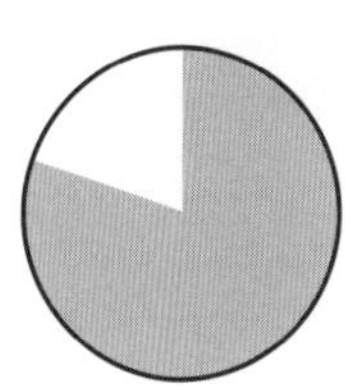 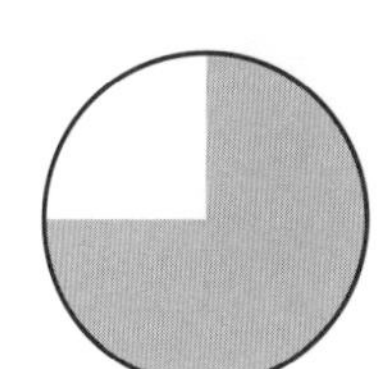

About 0% shaded	About 10% shaded	About 20% shaded	About 35% shaded	About 50% shaded	About 65% shaded	About 75% shaded	About 80% shaded	About 100% shaded

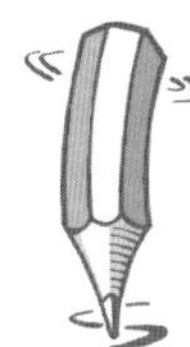

3. Estimate the percentage of each shape that is shaded.

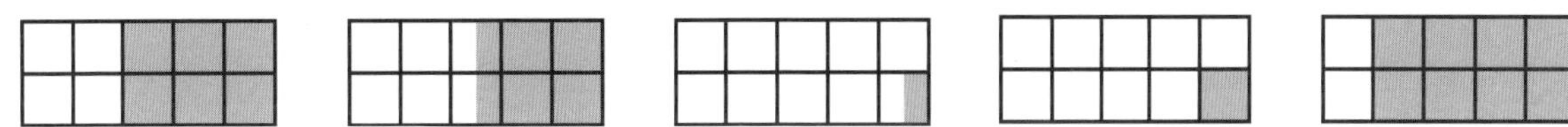

Finding percentages mentally 1

50% is $\frac{1}{2}$,   **25%** is $\frac{1}{4}$ and **75%** is $\frac{3}{4}$

Finding percentages in your head

There are several ways to find a percentage of a number or amount.

To find 50%: halve the number.
50% of **400** → **half** of **400** = **200**
50% of £**360** → **half** of £**360** = £**180**

To find 25%: halve the number and halve the answer.
25% of **400** → **half** of **400** = **200**,   → **half** of **200** = **100**
25% of £**360** → **half** of £**360** = £**180**, → **half** of £**180** = £**90**

To find 75%: halve the number and halve the answer,
then add your two answers together.
75% of **400** → **half** of **400** = **200**,   → **half** of **200** = **100**
                200 + **100** = **300**
75% of £**360** → **half** of £**360** = £**180**, → **half** of £**180** = £**90**
                £**180** + £**90** = £**270**

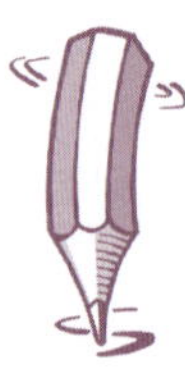

1. Find **50%** of:

a) 18 → _____   b) 24 → _____   c) 30 → _____   d) 72 → _____

e) 96 → _____   f) 150 → _____   g) 600 → _____   h) 1200 → _____

2. Find **25%** of:

a) 8 → _____   b) 16 → _____   c) 80 → _____   d) 48 → _____

e) 60 → _____   f) 100 → _____   g) 200 → _____   h) 880 → _____

3. Find **75%** of:

a) 4 → _____   b) 12 → _____   c) 48 → _____   d) 64 → _____

e) 52 → _____   f) 160 → _____   g) 360 → _____   h) 1000 → _____

Answers to Activities

Page 5

1. a) 2·**6** b) 6·**3** c) 7·**9**
   d) 13·**6** e) 28·**1** f) 56·**4**
   g) 8·**5**4 h) 9·**5**6 i) 23·**6**3

2. a) 2·7**2** b) 5·4**3** c) 9·0**2**
   d) 17·5**1** e) 65·9**7** f) 128·4**6**
   g) 5·4**9**2 h) 8·0**7**2 i) 19·6**0**8

3. a) 6 hundredths b) 4 tenths
   c) 9 tenths d) 4 hundredths
   e) 8 tenths f) 4 hundredths

Page 6

1. a) 0·67 b) 0·5 c) 0·8
   d) 0·61 e) 0·72 f) 0·9

2. a) 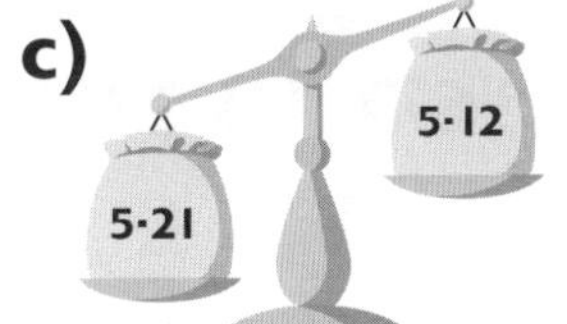

   b) 

   c) 

   d) 

   e)

   f)

Page 7

1. a) 0·56, 0·8, 0·85, 0·9
   b) 0·7, 0·75, 0·8, 0·83
   c) 0·38, 0·39, 0·4, 0·41
   d) 1·8, 1·86, 1·87, 1·9
   e) 1·98, 2·0, 2·07, 2·63

2. a) 0·42kg, 0·4kg, 0·30kg, 0·24kg
   b) 6·0m, 5·9m, 5·6m, 4·5m
   c) £9·50, £5·90, £0·59, £0·50
   d) 6·2kg, 6·15kg, 5·61kg, 1·65kg

Page 8

1. a) 1·2 b) 1·5 c) 1·9 d) 1·6

2. 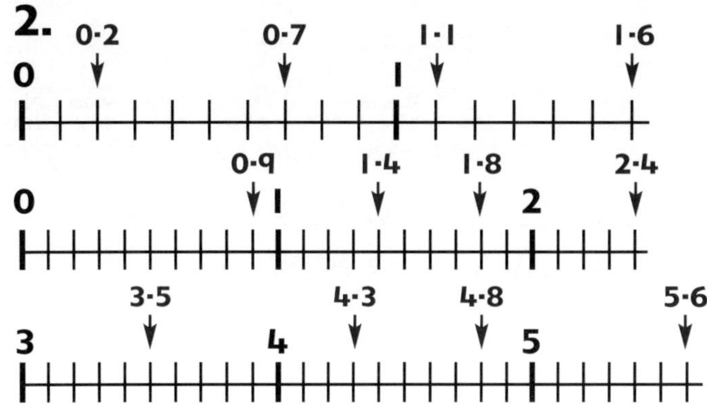

Page 9

1. a) 4 b) 4 c) 6
   d) 8 e) 11 f) 9
   g) 12 h) 13 i) 20
   j) 27 k) 31 l) 37

Page 11

$\frac{1}{6}$	0.8	$\frac{3}{5}$	0.5	$\frac{6}{10}$	$\frac{1}{10}$	0.1	0.34	$\frac{8}{9}$	0.1
$\frac{1}{2}$	$\frac{1}{5}$	$\frac{3}{8}$	$\frac{4}{5}$	0.4	$\frac{2}{5}$	0.73	$\frac{15}{100}$	0.76	0.63
$\frac{2}{7}$	0.5	0.1	0.3	0.2	$\frac{2}{3}$	$\frac{3}{7}$	0.15	$\frac{63}{100}$	$\frac{8}{9}$
0.12	$\frac{1}{4}$	0.25	$\frac{3}{10}$	$\frac{1}{7}$	$\frac{87}{100}$	$\frac{3}{5}$	0.75	0.11	0.2

Start … Finish

Page 12

1. a) 50% b) 80% c) 40%
   d) 30% e) 40% f) 15%

2. a) putting their all into it
   b) $\frac{1}{10}$ (10%) is fat
   c) all cds are half price

Page 13

1. 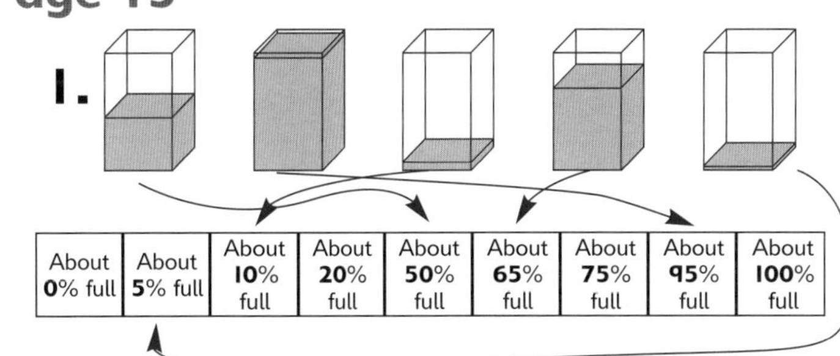

Answers to Activities

Page 13

2.
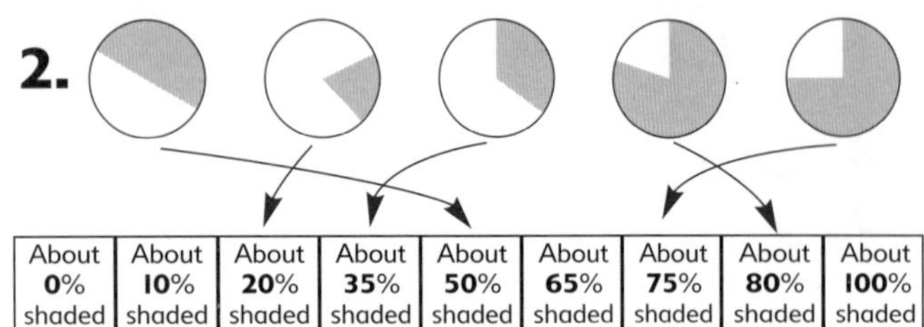

About 0% shaded	About 10% shaded	About 20% shaded	About 35% shaded	About 50% shaded	About 65% shaded	About 75% shaded	About 80% shaded	About 100% shaded

3. a) 60%   b) 50%   c) 5%
   d) 10%   e) 80%

Page 14

1. a) 9   b) 12   c) 15   d) 36
   e) 48   f) 75   g) 300   h) 600

2. a) 2   b) 4   c) 20   d) 12
   e) 15   f) 25   g) 50   h) 220

3. a) 3   b) 9   c) 36   d) 48
   e) 39   f) 120   g) 270   h) 750

Page 15

1. a) £16   b) 28g   c) 168m

2. a) £32   b) 100kg   c) 360m

3. a) £42   b) 480km   c) 72cm

4. a) £9   b) 27km   c) 30cm

5. a) £140   b) 7km   c) 175cm

Page 17

1.

$50\% = 0{\cdot}5 = \dfrac{1}{2}$ $80\% = 0{\cdot}8 = \dfrac{4}{5}$

$40\% = 0{\cdot}4 = \dfrac{2}{5}$ $30\% = 0{\cdot}3 = \dfrac{3}{10}$

$10\% = 0{\cdot}1 = \dfrac{1}{10}$ $20\% = 0{\cdot}2 = \dfrac{1}{5}$

$60\% = 0{\cdot}6 = \dfrac{3}{5}$ $90\% = 0{\cdot}9 = \dfrac{9}{10}$

$70\% = 0{\cdot}7 = \dfrac{7}{10}$ $1\% = 0{\cdot}01 = \dfrac{1}{100}$

2. a) $\dfrac{2}{5}$ b) $\dfrac{7}{5}$ c) 0·25   d) $\dfrac{9}{9}$

Page 18

1. a) 0·32   b) 0·27   c) 0·63
   d) 0·09   e) 0·01   f) 1 or 1·0

2. a) 72%   b) 45%   c) 87%
   d) 30%   e) 60%   f) 2%

3. a) $\dfrac{49}{100}$ b) $\dfrac{3}{20}$ c) $\dfrac{21}{25}$

   d) $\dfrac{1}{20}$ e) $\dfrac{1}{100}$ f) $\dfrac{3}{4}$

4. a) 30%   b) 90%   c) 80%
   d) 5%   e) 12%   f) 34%

Page 19

1. a) 4·8   b) 5·3   c) 6·1
   d) 8·6   e) 9·4   f) 10·4
   g) 14·9   h) 16·2   i) 21·0

2. a) 5, 4·8   b) 7, 7·5
   c) 13, 12·8   d) 23, 23·4
   e) 28, 27·7   f) 33, 32·9

Page 20

1. a) 3·9⑤   b) 4·③7   c) 8·9①
   d) 0·⑤7   e) 15·⓪3   f) 38·①7
   g) 78·⓪4   h) 10·③5

2. a) 3·45⑥   b) 5·80③
   c) 11·06⑨   d) 8·12③
   e) 35·50⑨1   f) 2·900⑦
   g) 18·46②3   h) 2015·23④5

3. a) 0·05   b) 0·9   c) 0·008
   d) 0·7   e) 0·006

Answers to Activities

Page 21

1. a) 0·24, 0·31, 0·5, 0·7, 0·86
   b) 0·594, 0·6, 0·65, 0·693, 0·72
   c) 3·278, 3·3, 3·467, 3·691, 3·72
   d) 6·007, 6·756, 7·063, 7·1, 7·6

2. a) £6.70, 660p, £6.50, £6.07
   b) 8·52kg, 6·85kg, 6·58kg,
      5·683kg
   c) 0·45m, 0·42m,
      0·348m, 0·3m
   d) 4·732m, 4·73m, 4·273m,
      4·237m,

3. 0·346, 0·364, 0·436, 0·463,
   0·634, 0·643, 3·046, 3·064,
   3·406, 3·460, 3·604, 3·640,
   4·036, 4·063, 4·306, 4·360,
   4·603, 4·630, 6·034, 6·043,
   6·304, 6·340, 6·403, 6·430,
   30·46, 30·64, 34·06, 34·60,
   36·04, 36·40, 40·36, 40·63,
   43·06, 43·60, 46·03, 46·30,
   60·34, 60·43, 63·04, 63·40,
   64·03, 64·30, 304·6, 306·4,
   340·6, 346·0, 360·4, 364·0,
   403·6, 406·3, 430·6, 436·0,
   460·3, 463·0, 603·4, 604·3,
   630·4, 634·0, 640·3, 643·0

Page 22

1. a) £76.80   b) £15.36
   c) £42.68   d) 29·40kg
   e) 120·96kg   f) 8·05kg
   g) 29·58ml   h) 145ml
   i) 250·10ml   j) 287·10m
   k) 42·48m   l) 132·48m

2. a) £9.45   b) £11.20

c) £39.33   d) £13.80
e) £33.60   f) £23.40

Page 23

1. 14 pairs of decimals adding to 1,
   in any order.

2. Any other pairs to add to 1

Page 24

1. a) 40·9   b) 83·9
   c) 92·5   d) 81·37
   e) 17·37   f) 20·94
   g) 20·06   h) 118·2

2. a) 14·2   b) 19·2
   c) 31·8   d) 18·48
   e) 279·9   f) 25·1
   g) 66·34   h) 83·46

Page 25

1. a) 12·5 litres
   b) 19·5 litres
   c) 122·4

2. a) 19 girls
   b) 6 games
   c) Dan = 75%, Ben = 74%

Answers to Tests

PROGRESS TEST 1 – Page 10

1. a) 7·**6**   b) 3·3**8**   c) 17·9**5** d) 24·**0**25

2. a) 4·7**6**   b) 154·7**3**
   c) 94·7**1**2   d) 20·0**2**4

3. a) 9 hundredths   b) 4 tenths
   c) 1 hundredth

4. a) 0·3   b) 0·79   c) 0·9

5. a) 0·59, 0·63, 0·66, 0·7

   b) 0·1, 0·18, 0·2, 0·81

   c) 0·49m, 0·5m, 0·58m, 0·61m

6. 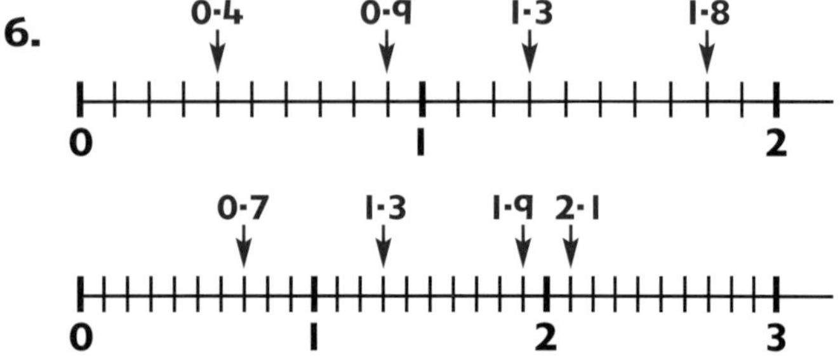

7. a) 4   b) 8   c) 25

Total marks = 28

PROGRESS TEST 2 – Page 16

1. $\frac{3}{4}$ joined to 0.75, $\frac{3}{10}$ joined to 0.3, and

   $\frac{4}{5}$ joined to 0.8

2. a) 50%   b) 30%   c) 35%

3. 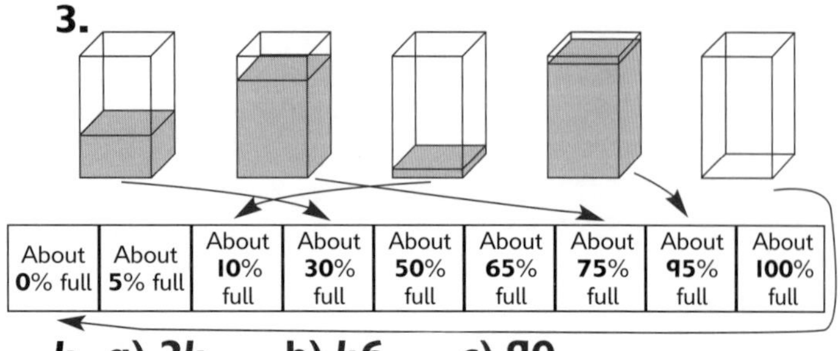

4. a) 24   b) 46   c) 90

5. a) 6   b) 15   c) 30

6. a) 6   b) 27   c) 300

7. a) £60   b) 45kg   c) 93m

8. a) £27   b) 90km   c) 225cm

Total marks = 26

FINAL TEST – Pages 26 to 28

1. a) 9·**67**   b) 8·**62**9   c) 15·**30**7 d) 124·**5**8

2. a) 4 hundredths   b) 9 tenths
   c) 6 hundredths   d) 4 hundredths

3. a) 0·23, 0·29, 0·3, 0·34
   b) 0·54m, 0·59m, 0·6m, 0·67m

4. 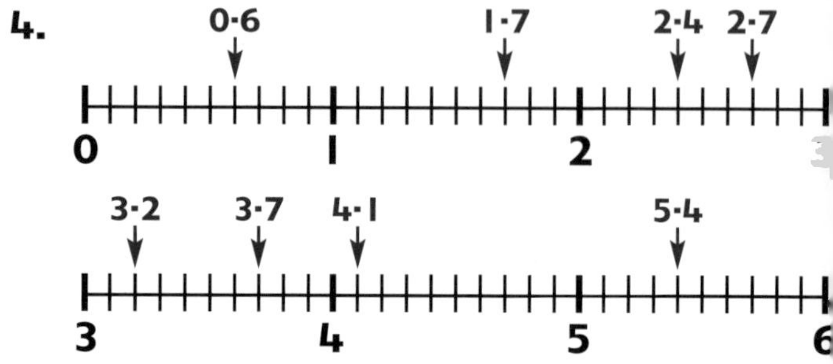

5. a) 5   b) 8   c) 36

6. $\frac{1}{4}$ joined to 0.25, $\frac{7}{10}$ joined to 0.7,

   and $\frac{1}{5}$ joined to 0.2

7. a) 18   b) 45   c) 225   d) 93

8. a) £140   b) 210kg   c) 14m   d) 84

9. a) £4.50   b) 12km   c) 60cm   d) 24g

10.

$30\% = 0.3 = \frac{3}{10}$ $40\% = 0.4 = \frac{2}{5}$

$70\% = 0.7 = \frac{7}{10}$ $80\% = 0.8 = \frac{4}{5}$

$10\% = 0.1 = \frac{1}{10}$ $60\% = 0.6 = \frac{3}{5}$

11. a) 0·15   b) 0·49   c) 0·06   d) 0·78

12. a) 32%   b) 40%   c) 7%   d) 51%

13. a) $\frac{39}{100}$ b) $\frac{4}{25}$ c) $\frac{1}{50}$ d) $\frac{13}{20}$

14. a) 70%   b) 24%   c) 90%   d) 95%

15. a) 4·6   b) 5·5   c) 37·0

16. a) 8·45**9**   b) 31·43**7**
    c) 12·02**98**   d) 1024·45**81**

17. a) 46·4   b) £8·64   c) 141·12m

18. a) 85·8   b) 124·2   c) 27·4   d) 18·79

19. a) 57·66   b) 193·25
    c) 217·28   d) 11·312

20. a) 107·8   b) 25 girls

Total marks = 84

Finding other percentages in your head

To find other percentages in your head you can use **10%**:

To find 10%: divide the number by 10
10% of **400** → **400** ÷ 10 = **40**
10% of **£360** → **£360** ÷ 10 = **£36**

To find 20%: divide by 10 and double
20% of **400** → **400** ÷ 10 = **40**,   40 × 2 = 80
20% of **£360** → **£360** ÷ 10 = **£36**   £36 × 2 = £72

To find 30%: divide by 10 and multiply by 3
30% of **400** → **400** ÷ 10 = **40**,   40 × 3 = 120
30% of **£360** → **£360** ÷ 10 = **£36**   £36 × 3 = £108

To find 15%, first find 10% then halve your answer to get 5% and add the two answers together. 15% of 60 = 6 + 3 = 9

To find **40%**: divide by **10** and multiply by **4**
For **all percentages that are multiples of 10**, use your answer to **10%** to help you.

1. Find **10%** of:

   a) £160 → _______   **b)** 280g → _______   **c)** 1680m → _______

2. Find **40%** of:

   a) £80 → _______   **b)** 250kg → _______   **c)** 900m → _______

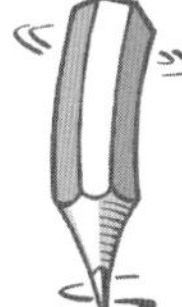

3. Find **60%** of:

   a) £70 → _______   **b)** 800km → _______   **c)** 120cm → _______

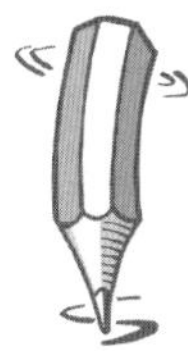

4. Find **15%** of:

   a) £60 → _______   **b)** 180km → _______   **c)** 200cm → _______

5. Find **35%** of:

   a) £400 → _______   **b)** 20km → _______   **c)** 500cm → _______

Progress Test 2

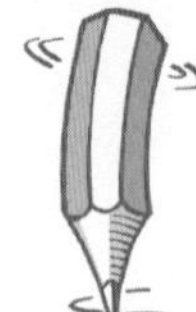

1. Join pairs of decimals and fractions with the same value.

$\dfrac{3}{4}$ **0.3** $\dfrac{4}{5}$ **0.4**   **0.75** $\dfrac{3}{10}$ **0.8**

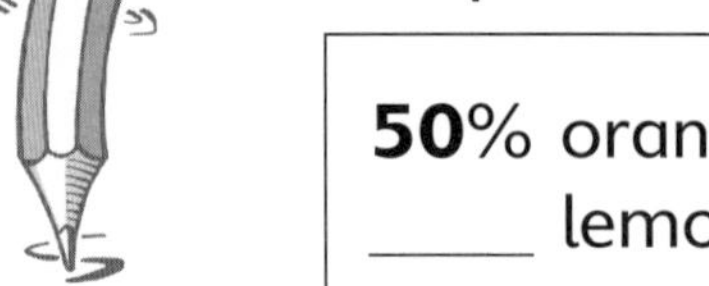

2. Complete these food labels, making sure they total **100%**.

50% oranges
_____ lemons

65% beans
5% salt
_____ tomatoes

55% potatoes
_____ cheese
10% water

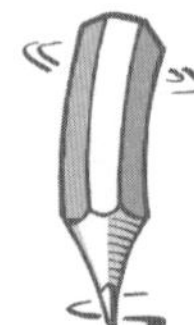

3. Match each container with an appropriate estimate.

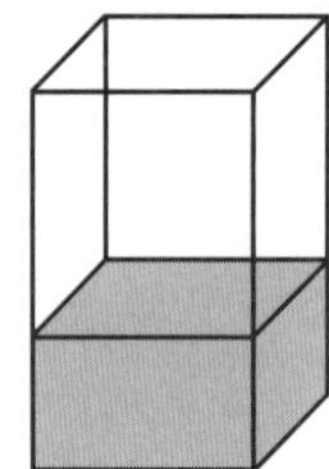 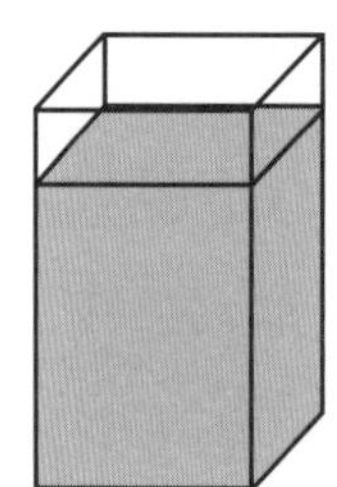 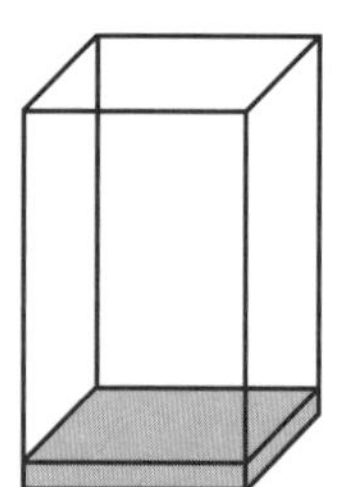 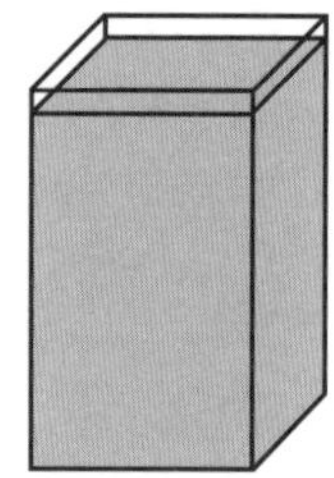 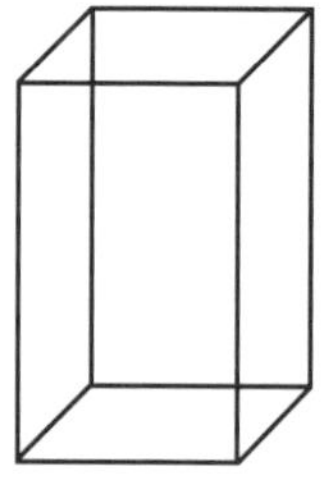

About 0% full	About 5% full	About 10% full	About 30% full	About 50% full	About 65% full	About 75% full	About 95% full	About 100% full

4. Find **50%** of:

a) 48 → _________ b) 92 → _________ c) 180 → _________

5. Find **25%** of:

a) 24 → _________ b) 60 → _________ c) 120 → _________

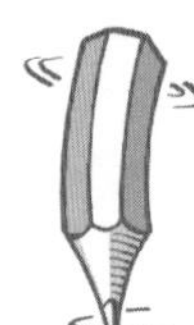

6. Find **75%** of:

a) 8 → _________ b) 36 → _________ c) 400 → _________

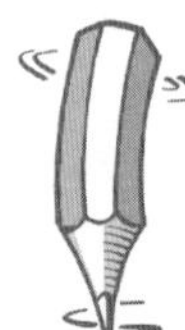

7. Find **30%** of:

a) £200 → _______ b) 150kg → _______ c) 310m → _______

8. Find **45%** of:

a) £60 → _______ b) 200km → _______ c) 500cm → _______

Do you remember on page 11 we said that decimals and fractions are like two different languages? We can say the same thing in different ways. Well, percentages are another language. They are another way of saying the same thing.

Look at this container. We can write:

(Fractions)   It is $\frac{1}{2}$ full

(Decimals)   It is **0·5** full
(Percentages) It is **50%** full

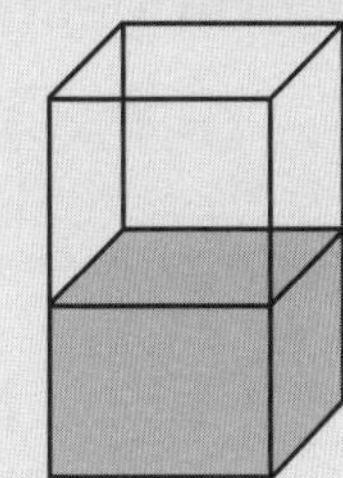

Percentages, fractions and decimals like these all have the same value:

60%, $\frac{3}{5}$ and **0·6**   **90%**, $\frac{9}{10}$ and **0·9**

We can place them all on a number line, like this

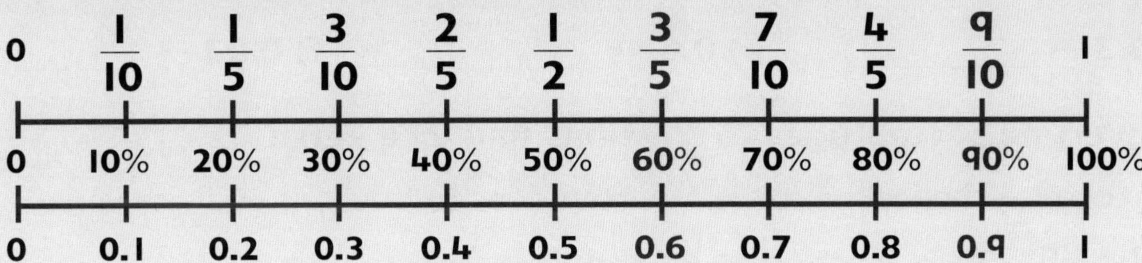

1. Use the number lines to help you to complete this table.

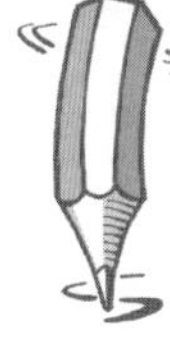

Percentage	Decimal	Fraction
50%		
40%		
	0·1	
	0·6	
		$\frac{7}{10}$

Percentage	Decimal	Fraction
80%		
	0·3	
		$\frac{1}{5}$
	0·9	
	0·01	$\frac{1}{100}$

2. Draw a ring around the odd one out in each set.

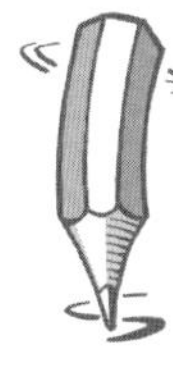

a) 25%   0·25 $\frac{2}{5}$

b) 75%   0·75 $\frac{7}{5}$

c) 40% $\frac{2}{5}$ 0·25

d) 90%   0·9 $\frac{9}{9}$

Converting percentages to decimals and vice versa

We can change other percentages into decimals like this:

just divide by **100**   48% → 48 ÷ 100 = 0·48   65% → 65 ÷ 100 = 0·65

To change decimals into percentages:

just multiply by **100**   0·3 → 0·3 × 100 = 30%   0·78 → 0·78 × 100 = 78%

1. Change these percentages into decimals.

   a) 32% → _________   b) 27% → _________   c) 63% → _________

   d) 9% → _________   e) 1% → _________   f) 100% → _________

2. Change these decimals into percentages.

   a) 0.72 → _________   b) 0.45 → _________   c) 0.87 → _________

   d) 0.3 → _________   e) 0.6 → _________   f) 0.02 → _________

Converting percentages to fractions and vice versa

We can also change, or convert, percentages to fractions and vice versa.

Write 48% as a fraction in its simplest form:

Write the percentage as a fraction with a denominator of **100**. Then divide the numerator and denominator by the same number to find the simplest form:

$$48\% = \frac{48}{100} \frac{\div\ 4}{\div\ 4} = \frac{12}{25}$$

Change $\frac{17}{20}$ **into a percentage:**

Convert $\frac{17}{20}$ into a fraction with a denominator of **100**: $\frac{17}{20} \frac{\times\ 5}{\times\ 5} = \frac{85}{100} = 85\%$

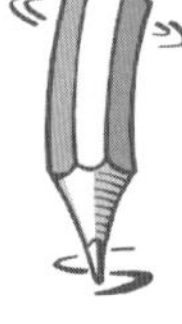

3. Change these percentages into fractions in their simplest forms.

   a) 49% → _________   b) 15% → _________   c) 84% → _________

   d) 5% → _________   e) 1% → _________   f) 75% → _________

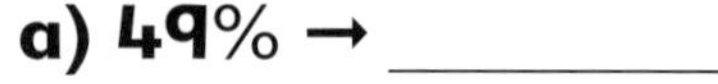

4. Change these fractions into percentages.

   a) $\frac{3}{10}$ → _________   b) $\frac{9}{10}$ → _________   c) $\frac{4}{5}$ → _________

   d) $\frac{1}{20}$ → _________   e) $\frac{3}{25}$ → _________   f) $\frac{17}{50}$ → _________

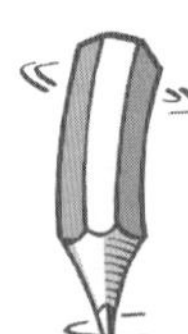

Rounding decimals to the nearest tenth

We can round decimals to the nearest tenth in the same way as the rounding to the nearest whole number we did on page **9**.

Look at this question: **Round 0·37 to the nearest tenth.**

One way of rounding is to look at the number on a number line and see which tenth the number is nearer to.

0·3   0·37   0·4

0·37 is nearer to **0·4**

Here is another way...

- Firstly see what you are rounding to.   Round **0·37** to the nearest **tenth**.
- Then point to this digit in the number...   U   t   h
                                              0 · 3   7

- Now look at the digit to the right of it... 0 · 3   7
- If it's smaller than **5** the tenths digit stays the same. If it's **5** or larger the tenths digit must go up one.

          0·37 rounded to the nearest tenth is **0·4**

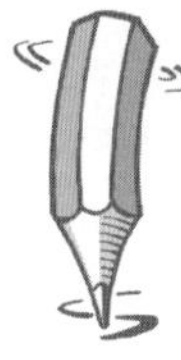

1. Round these numbers to the nearest tenth.

a) (4·76) → ( )   b) (5·28) → ( )   c) (6·13) → ( )

d) (8·64) → ( )   e) (9·42) → ( )   f) (10·39) → ( )

g) (14·91) → ( )   h) (16·15) → ( )   i) (20·95) → ( )

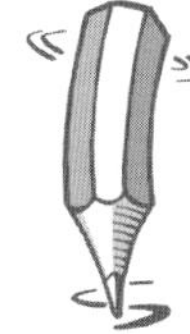

2. Round these numbers to the nearest whole number and to the nearest tenth.

	nearest whole number	nearest tenth
a) 4·83	5	4.8
c) 12·76		
e) 27·725		

	nearest whole number	nearest tenth
b) 7·49		
d) 23·41		
f) 32·85		

Tenths, hundredths and thousandths

Tenths and hundredths

On page **5** we learnt that the first digit to the right of the decimal point is the **tenths** digit.

The second digit to the right of the decimal point is the **hundredths** digit.

Units	·	tenths	hundredths
6	·	8	5

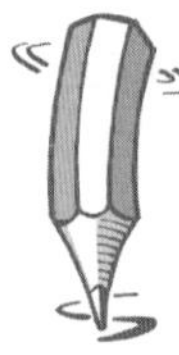

I. Draw a ring round the **tenths** digit and underline the <u>hundredths</u> digit in these numbers.

a) 3·95   b) 4·3 7   c) 8·9 1   d) 0·5 7

e) 1 5·0 3   f) 3 8·1 7   g) 7 8·0 4   h) 1 0·3 5

Thousandths

The third digit to the right of the decimal point is the **thousandths** digit.

10 thousandths make **1** hundredth and **1000** thousandths make a whole one.

Units	·	tenths	hundredths	**thousandths**
6	·	4	8	7

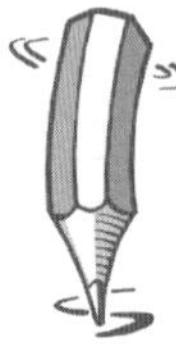

2. Draw a ring round the thousandths digit in these numbers.

a) 3·4 5 6   b) 5·8 0 3   c) 1 1·0 6 9   d) 8·1 2 3

e) 3 5·5 0 9 1   f) 2·9 0 0 7   g) 1 8·4 6 2 3   h) 2 0 1 5·2 3 4 5

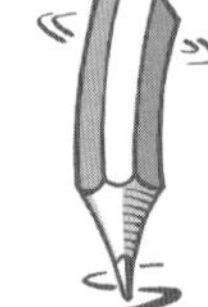

3. Colour the brick that shows the value of the underlined digit.

a) 8.3<u>5</u>6	50	5·0	0·5	0·05	0·005
b) 6.<u>9</u>03	90	9·0	0·9	0·09	0·009
c) 9.51<u>8</u>	80	8·0	0·8	0·08	0·008
d) 12.<u>7</u>04	70	7·0	0·7	0·07	0·007
e) 23.82<u>6</u>	60	6·0	0·6	0·06	0·006

On page **7** we learnt how to order decimals with up to two digits after the decimal point.
We can easily order decimals with more than two digits after the point.

Put these decimals in order of size, largest first **0·783 0·88 0·78 0·9 0·792**

Which number has most tenths? ⟶ **0·9**

Which has the next greatest number of tenths? ⟶ **0·88**

Which numbers have the next greatest number of tenths? ⟶
0.783, 0.78 or **0.792**

Which of these has most hundredths? ⟶ **0·792**

Which numbers have the next greatest number of hundredths? **0·783** or **0·78**

Which of these has more thousandths? ⟶ **0·783**

Which is the smallest number? ⟶ **0·78**

From largest to smallest the order is: **0·9, 0·88, 0·792, 0·783, 0·78**

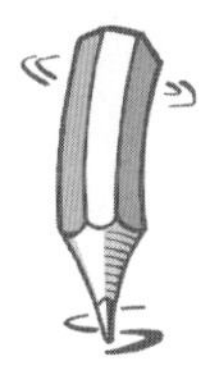

1. Put these decimals in order of size, **smallest** first.

a) **0·7, 0·24, 0·31, 0·86, 0·5** _______________________________

b) **0·6, 0·65, 0·594, 0·72, 0·693** _______________________________

c) **3·72, 3·278, 3·691, 3·3, 3·467** _______________________________

d) **7·063, 7·6, 6·756, 6·007, 7·1** _______________________________

2. Write these in order of size, **largest** first.

a) **£6·07, £6·70, 660p, £6·50**

b) **5·683kg, 6·58kg, 6·85kg, 8·52kg**

c) **0·348m, 0·3m, 0·45m, 0·42m**

d) **4·327m, 4·273m, 4·732m, 4·73m**

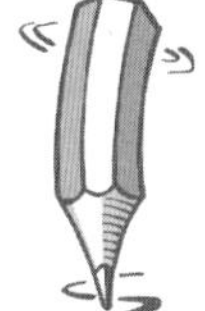

3. Use these cards to make as many different numbers as you can.
Write your numbers in order, **smallest** first.

6 3 4 . 0

There are different ways of calculating percentages on a calculator because a percentage can be written as either a fraction or a decimal.

The fraction way:

Think of a percentage as a fraction 'out of **100**' or 'divided by **100**'.

To find **48**% of **56** on a calculator, we can key in **48**% as a fraction

$$\frac{48}{100} \times 56 = 26\cdot88 \quad \longleftarrow \quad \text{Key into the calculator}$$

The decimal way:

Another way is to write each percentage as a decimal.

To find **48**% of **56** on a calculator, we can key in **48**% as a decimal.

$$0\cdot48 \times 56 = 26\cdot88$$

Choose which way you like best but always approximate first, like this: **48**% is a bit less than a half.

$$48\% \text{ of } 56 = 26\cdot88 \longrightarrow \frac{1}{2} \text{ of } 60 = 30.$$

26·88 is a bit less than **30** so your answer is likely to be correct.

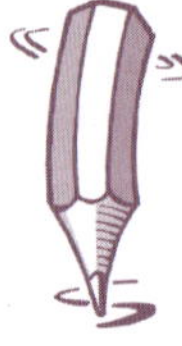

1. Use a calculator to find these percentages. Approximate first.

a) **64**% of £**120** = _______   b) **32**% of £**48** = _______

c) **44**% of £**97** = _______   d) **21**% of **140**kg = _______

e) **54**% of **224**kg = _______   f) **23**% of **35**kg = _______

g) **34**% of **87**ml = _______   h) **58**% of **250**ml = _______

i) **41**% of **610**ml = _______   j) **99**% of **290**m = _______

k) **36**% of **118**m = _______   l) **72**% of **184**m = _______

2. Work out these percentages to find the sale prices. Approximate first.

a) **45**% of £**21**   b) **32**% of £**35**   c) **57**% of £**69**

_______   _______   _______

d) **23**% of £**60**   e) **84**% of £**40**   f) **39**% of £**60**

_______   _______   _______

Decimals with a total of one

Pairs of decimals

In the same way that we can find pairs of numbers that add to **10**

$$7 + 3 = 10$$

or pairs with a total of **100**,
we can find pairs of decimals that add to **1**

$$63 + 37 = 100$$
$$0·64 + 0·36 = 1$$

Remember 10 tenths make one whole or 100 hundredths make one whole.

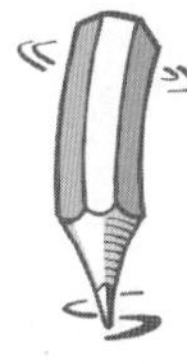

1. Find pairs of decimals with a total of **1** from the grid and write them below.

0·7	0·22	0·34	0·2
0·01	0·44	0·72	0·92
0·9	0·17	0·39	0·78
0·81	0·6	0·13	0·3
0·77	0·97	0·28	0·23
0·66	0·51	0·11	0·38
0·08	0·4	0·49	0·19
0·65	0·83	0·47	0·1
0·02	0·8	0·99	0·35

$0·7 + 0·3 = 1$ ___ + ___ = 1

___ + ___ = 1   ___ + ___ = 1

___ + ___ = 1   ___ + ___ = 1

___ + ___ = 1   ___ + ___ = 1

___ + ___ = 1   ___ + ___ = 1

___ + ___ = 1   ___ + ___ = 1

___ + ___ = 1   ___ + ___ = 1

2. Find ten more pairs of decimals with a total of **1** and write them below.

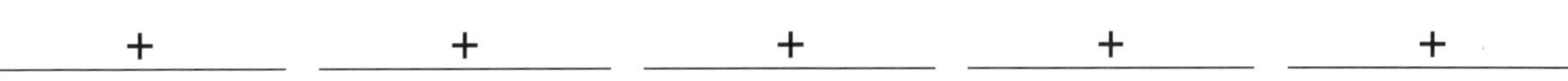

_____ + _____   _____ + _____   _____ + _____   _____ + _____   _____ + _____

_____ + _____   _____ + _____   _____ + _____   _____ + _____   _____ + _____

Adding and subtracting decimals

We can add decimals in the same way we add whole numbers. Just make sure to line up the decimal points.

46·6 + 19·2

T	U · t	Approx. 50 + 20 = 70
	4	6 · 6
+	1	9 · 2
	6	5 · 8

32·74 + 23·69

T	U · t	h	Approx. 30 + 20 = 50
	3	2 · 7	4
+	2	3 · 6	9
	5	6 · 4	3

When you are adding decimals it is very important to get an approximate answer first. This helps you to be sure that the answer isn't **6·58** or **564·3**

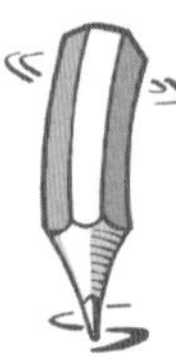

1. Add these numbers.

a)
```
  1 6·7
+ 2 4·2
_______
```

b)
```
  4 5·6
+ 3 8·3
_______
```

c)
```
  5 2·7
+ 3 9·8
_______
```

d)
```
  4 3·5 4
+ 3 7·8 3
_________
```

e)
```
  1 2·8 2
+   4·5 5
_________
```

f)
```
  1 2·5 3
+   8·4 1
_________
```

g)
```
  1 1·6 4
+   8·4 2
_________
```

h)
```
  8 2·3 2
+ 3 5·8 8
_________
```

We can subtract decimals in the same way we subtract whole numbers. Just make sure to line up the decimal points.

63·7 – 31·2

T	U · t	Approx. 60 – 30 = 30
	6	3 · 7
–	3	1 · 2
	3	2 · 5

87·5 – 34·72

T	U · t	h	Approx. 90 – 30 = 60
	8	7 · 5	0
–	3	4 · 7	2
	5	2 · 7	8

Tip: Fill any blank columns with a **0** to help line them up.

Check your answer by **adding** the bottom two lines of the calculations.

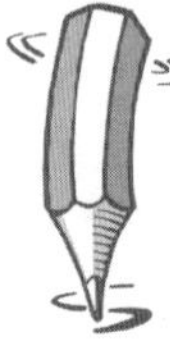

2. Subtract these numbers.

a)
```
  2 8·7
– 1 4·5
_______
```

b)
```
  3 6·5
– 1 7·3
_______
```

c)
```
  7 5·6
– 4 3·8
_______
```

d)
```
  6 7·8 2
– 4 9·3 4
_________
```

e)
```
  3 7 8·6
–   9 8·7
_________
```

f)
```
  2 1 4·5
– 1 8 9·4
_________
```

g)
```
  1 3 6·0 4
–   6 9·7
_________
```

h)
```
  1 3 2·8
–   4 9·3 4
_________
```

Word problems

Word problems

When faced with a problem, follow these steps:

- **read the problem carefully**
- **look for any useful words in the question**
- **write down any important numbers in the question**
- **decide what operations to use**
- **get an approximate answer**
- **decide whether to use a written or mental method and work it out**
- **finally check your answer**

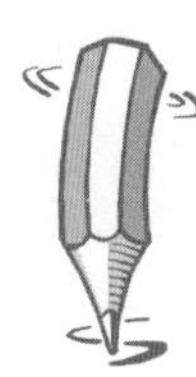

1. Solve these word problems. You can use a calculator, or scrap paper.

a) There are **87·5** litres of water in **7** fish tanks. There is the same amount of water in each tank.
How much water is in each tank? _________________________ litres

b) In **5** visits to the garage John puts a total of **97·5** litres of petrol into his car.
How much does he put in on average? _________________________ litres

c) Jack thinks of a decimal. He divides it by **9**.
His answer is **13·6**
What was the number he thought of? _________________________

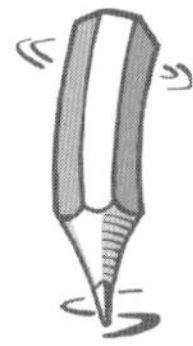

2. Solve these word problems about percentages. You can use a calculator, or scrap paper.

a) A school party of **50** is on the London Eye.
52% are boys, **10**% are adults.
How many are girls? _________________________ girls

b) The school hockey team played **15** games.
They won **60**% of them.
How many games did they not win? _________________________ games

c) Dan scored **30** out of **40** in a test.
Ben scored **148** out of **200** in another test.
What percentage did they each score?

Dan scored __________ % and Ben scored __________ %

1. Draw a ring round the tenths digit and underline the **hundredths** digit in these numbers.

a) 9·6 7   b) 8·6 2 9   c) 1 5·3 0 7   d) 1 2 4·5 8

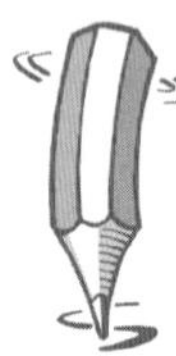

2. What is the pink digit worth in each of these numbers?

a) 7·5 4   b) 3 2·9 3   c) 5 8·2 6 8   d) 2 1 6·2 4 5

_______   _______   _______   _______

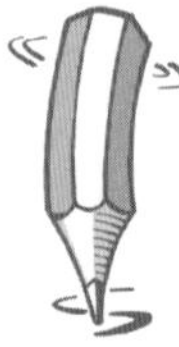

3. Put these decimals in order of size, **smallest** first.

a) 0·3, 0·23, 0·34, 0·29 ______________________________

b) 0·67m, 0·54m, 0·6m, 0·59m _______________________

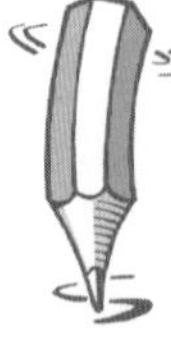

4. Mark these numbers on the number lines.

a) 0·6, 2·4, 1·7, 2·7

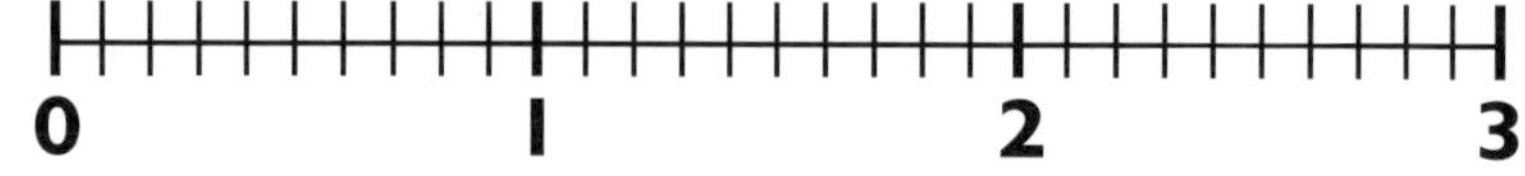

b) 5·4, 3·7, 4·1, 3·2

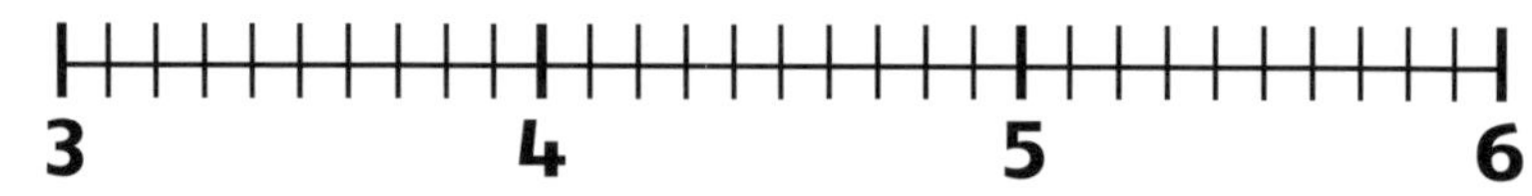

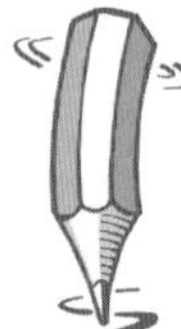

5. Round these numbers to the nearest whole numbers.

a) 4·5 → ◯   b) 7·51 → ◯   c) 36·302 → ◯

6. Join pairs of decimals and fractions with the same value.

| $\frac{1}{4}$ | 0.7 | $\frac{1}{5}$ | 0.9 | 0.25 | $\frac{7}{10}$ | 0.2 |

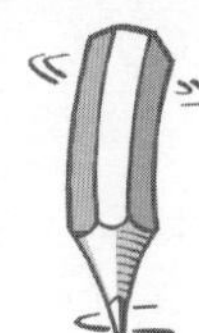

7. Without a calculator find **75%** of:

a) 24 → _____ b) 60 → _____ c) 300 → _____ d) 124 → _____

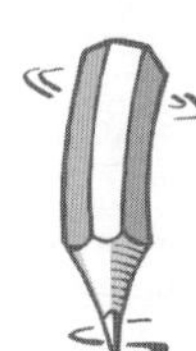

8. Without a calculator find **70%** of:

a) £200 → _____ b) 300kg → _____ c) 20m → _____ d) 120 → _____

9. Without a calculator find **15%** of:

a) £30 → _____ b) 80km → _____

c) 400cm → _____ d) 160g → _____

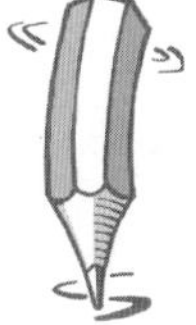

10. Complete this table.

Percentage	Decimal	Fraction
30%		
	0·7	
		$\frac{1}{10}$

Percentage	Decimal	Fraction
40%		
	0·8	
		$\frac{3}{5}$

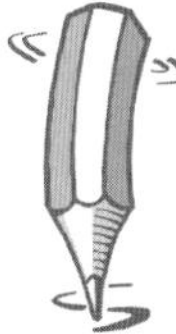

11. Change these percentages into decimals.

a) 15% → _____ b) 49% → _____ c) 6% → _____ d) 78% → _____

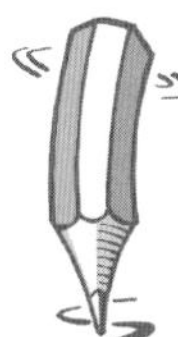

12. Change these decimals into percentages.

a) 0·32 → _____ b) 0·4 → _____ c) 0·07 → _____ d) 0·51 → _____

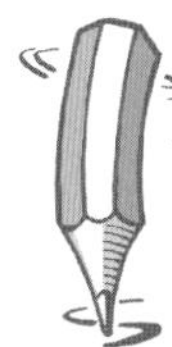

13. Change these percentages into fractions in their simplest form.

a) 39% → _____ b) 16% → _____ c) 2% → _____ d) 65% → _____

14. Change these fractions into percentages.

a) $\dfrac{7}{10}$ → _____ b) $\dfrac{6}{25}$ → _____ c) $\dfrac{45}{50}$ → _____ d) $\dfrac{19}{20}$ → _____

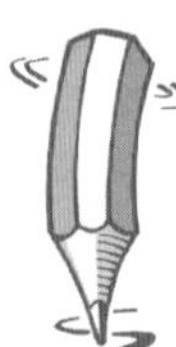

15. Round these numbers to the nearest tenth.

a) (4·57) → ( ) b) (5·49) → ( ) c) (36·95) → ( )

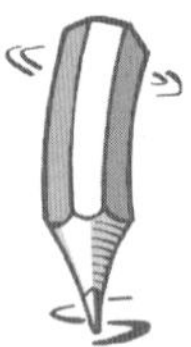

16. Draw a ring round the thousandths digit in these numbers.

a) 8·4 5 9 b) 3 1·4 3 7 c) 1 2·0 2 9 8 d) 1 0 2 4·4 5 8

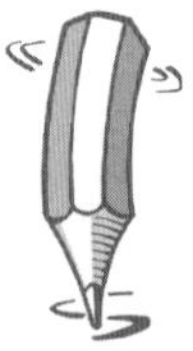

17. Use a calculator to find these percentages. Approximate first.

a) **58**% of **80** = _____ b) **12**% of **£72** = _____ c) **84**% of **168**m = _____

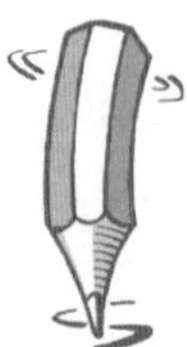

18. Without using a calculator, answer these questions.

a)
```
  2 7·6
+ 5 8·2
───────
```

b)
```
  6 5·4
+ 5 8·8
───────
```

c)
```
  7 6·7
− 4 9·3
───────
```

d)
```
  8 3·4 1
− 6 4·6 2
─────────
```

19. Without using a calculator, answer these questions.

a) 38·5 + 19·16 __________ b) 125·45 + 67·8 __________

c) 235·18 − 17·9 __________ d) 29·34 − 18·028 __________

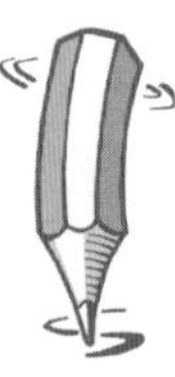

20. Solve these word problems. You can use a calculator.

a) Jane thinks of a decimal. She divides it by **7**.
   Her answer is **15·4**
   What was the number she thought of? __________

b) A school party of **50** visits a museum.
   42% are boys, **8**% are adults.
   How many are girls? __________ girls